Level 1

BRAINY SCIENCE READERS

DO YOU KNOW QUANTUM PHYSICS?

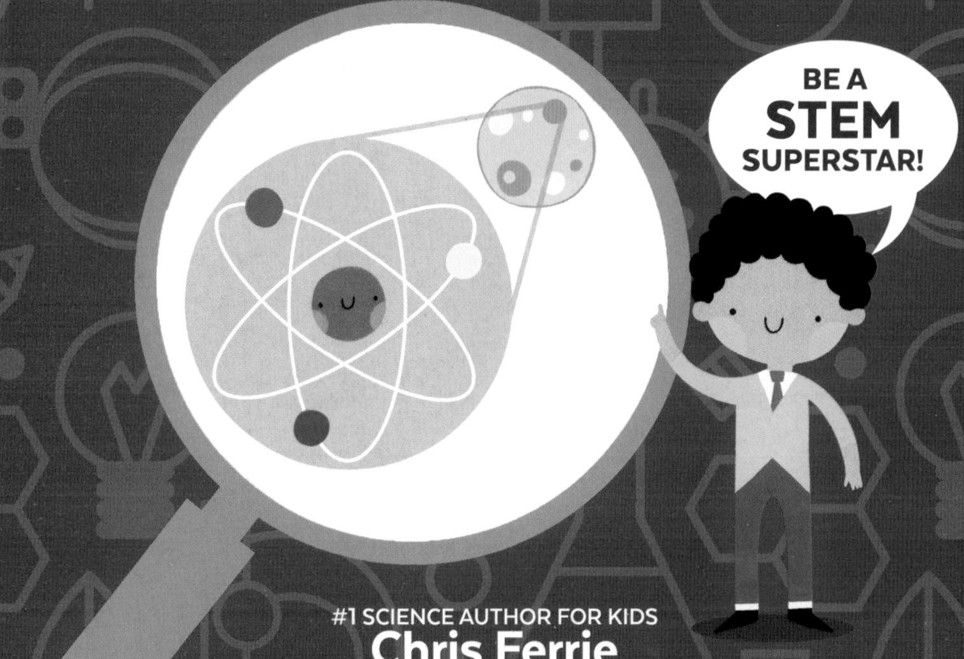

#1 SCIENCE AUTHOR FOR KIDS
Chris Ferrie

sourcebooks eXplore

ALSO AVAILABLE

Text adapted by Brooke Vitale
Illustrations by Chris Ferrie and Lindsay Dale-Scott

All art was created digitally.

Published by Sourcebooks eXplore, an imprint of Sourcebooks Kids
P.O. Box 4410, Naperville, Illinois 60567-4410
(630) 961-3900
sourcebookskids.com

This product conforms to all applicable CPSC and CPSIA standards.

Cataloging-in-Publication Data is on file with the Library of Congress.

Source of Production: Wing King Tong Paper Products Co. Ltd., Shenzhen, Guangdong Province, China
Date of Production: May 2022
Run Number: 5025963

Printed and bound in China.
WKT 10 9 8 7 6 5 4 3 2 1

Level 1

Dear Grown-up:

Welcome to the wonderful world of Brainy Science! Our mission is to help kids take their first steps to becoming independent readers! BRAINY SCIENCE will improve reading skills while immersing children into scientific theory. So blast off with BRAINY SCIENCE, and observe as your budding scientist learns to READ and draw their own conclusions!

The Brainy Reading Method

Level 1

Level 1: Beginner Reader
Pre K–Grade 1
Easy vocabulary. Short sentences. Word repetition. Simple content and stories. Correlation between art and text.

Level 2

Level 2: Emerging Reader
Kindergarten–Grade 2
Letter blends. Compound sentences. Contractions.
Simple, high-interest storylines. Art offers visual cues to decipher text.

Level 3

Level 3: Reading Alone
Grade 1–Grade 3
Longer, more complex storylines. Story told in paragraph form. Character development. More challenging letter blends and multisyllable words.
Art enhances the story.

Hi there, big thinker.
Are you ready to learn?

MARIE CURIE

Today we will talk about Quantum Physics.

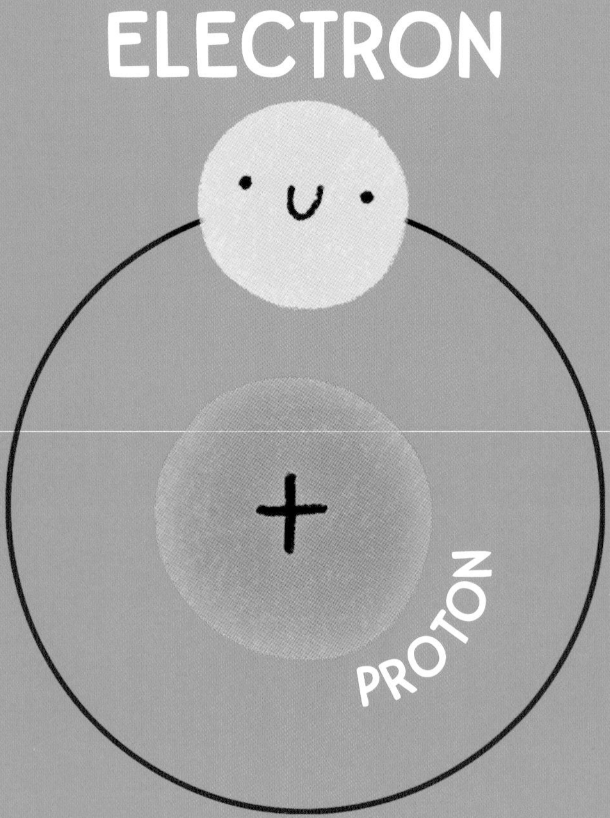

We will learn about the parts of an atom.

We will talk about
how the parts move.
Are you ready?
Let us begin.

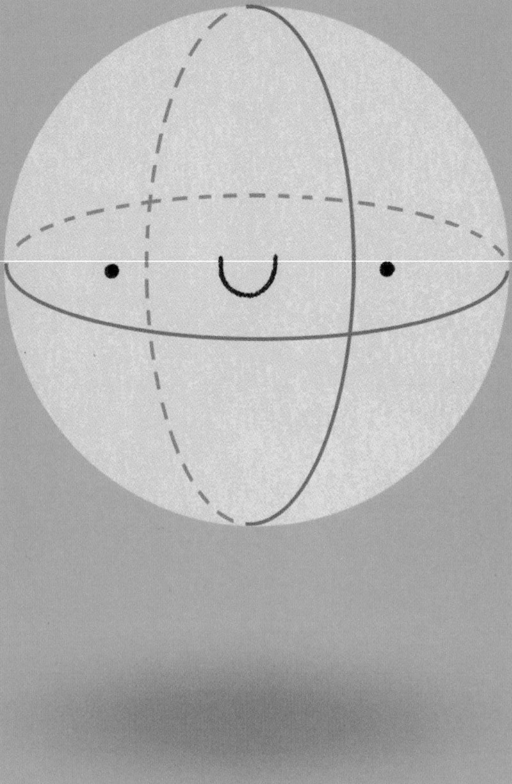

This is a ball.
The ball is not moving.

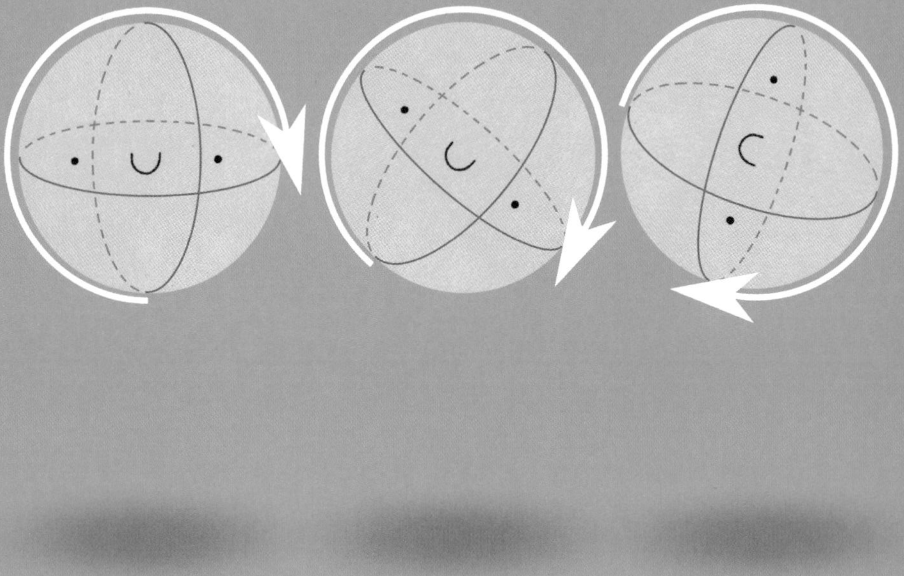

This ball is moving.
It is rolling.

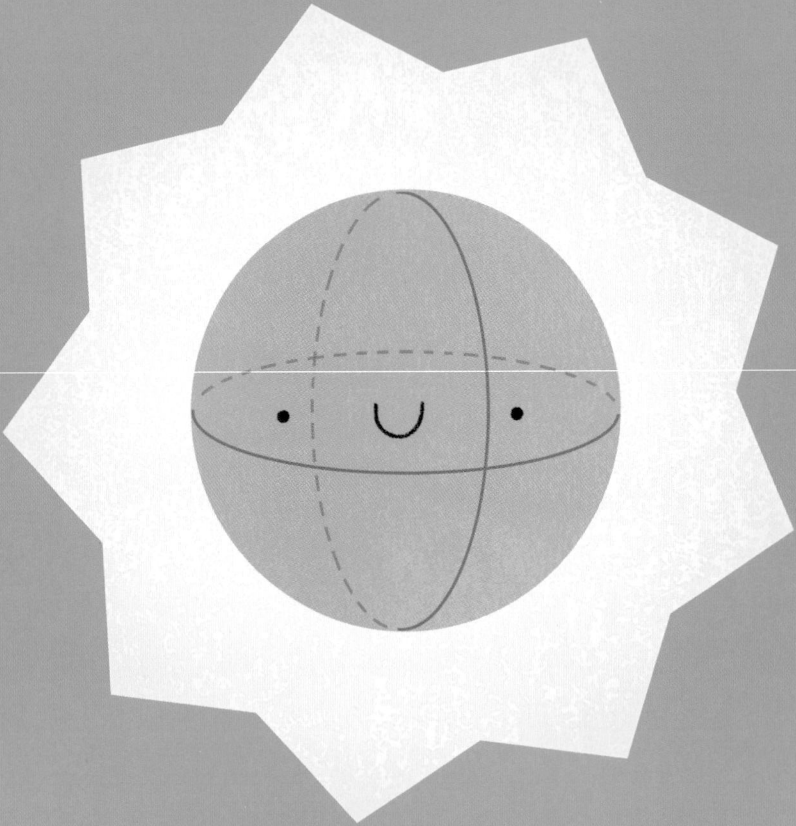

This ball has energy. Energy is how things change and move.

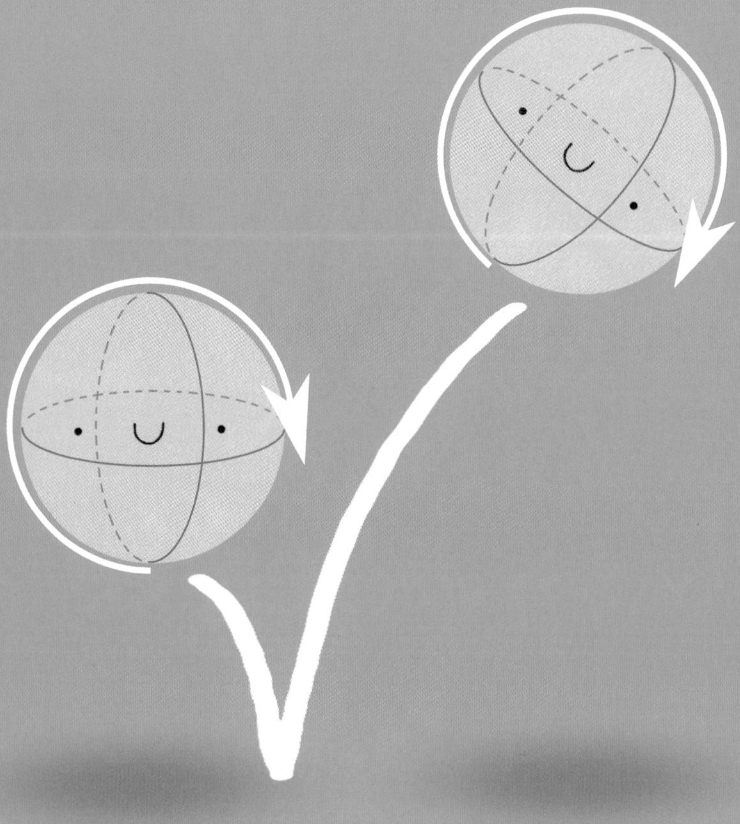

Energy is what makes the ball roll.

It is what makes the ball bounce.

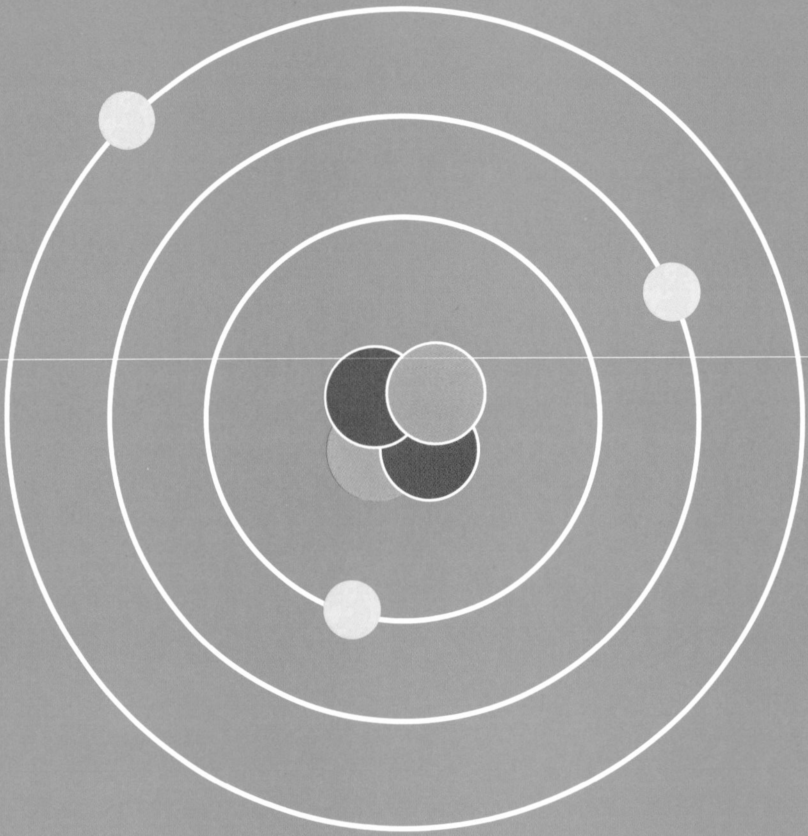

The ball is made from
lots of smaller parts.
Those parts are called atoms.

We cannot see atoms
with our eyes.
They are too small.

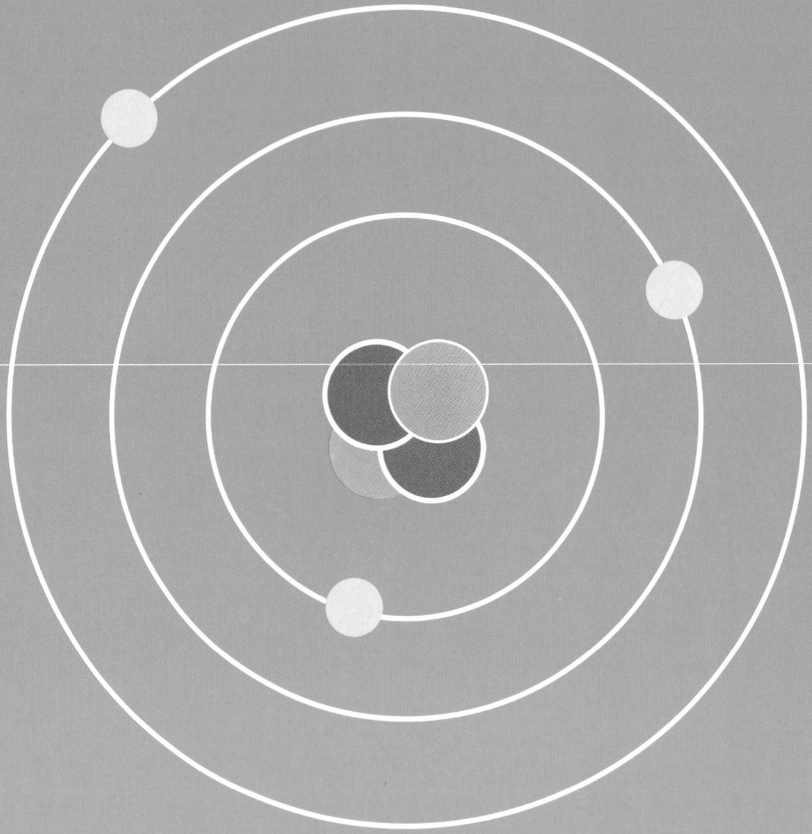

Atoms are made up of three parts. One part is called a neutron. It lives in the middle of the atom.

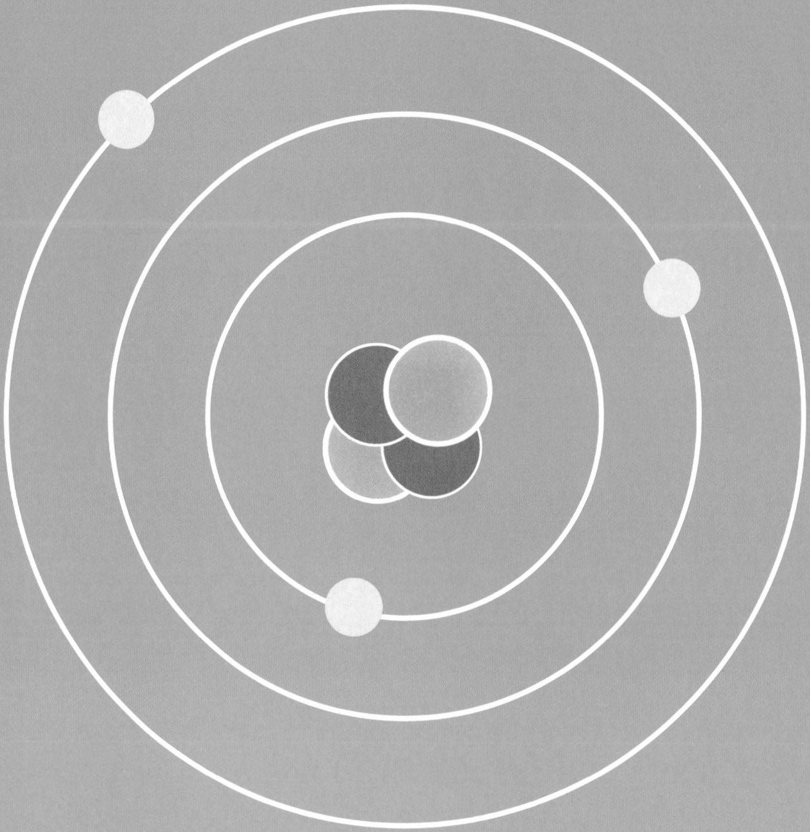

One part is called a proton.
It lives in the middle too.
The middle is called the nucleus.

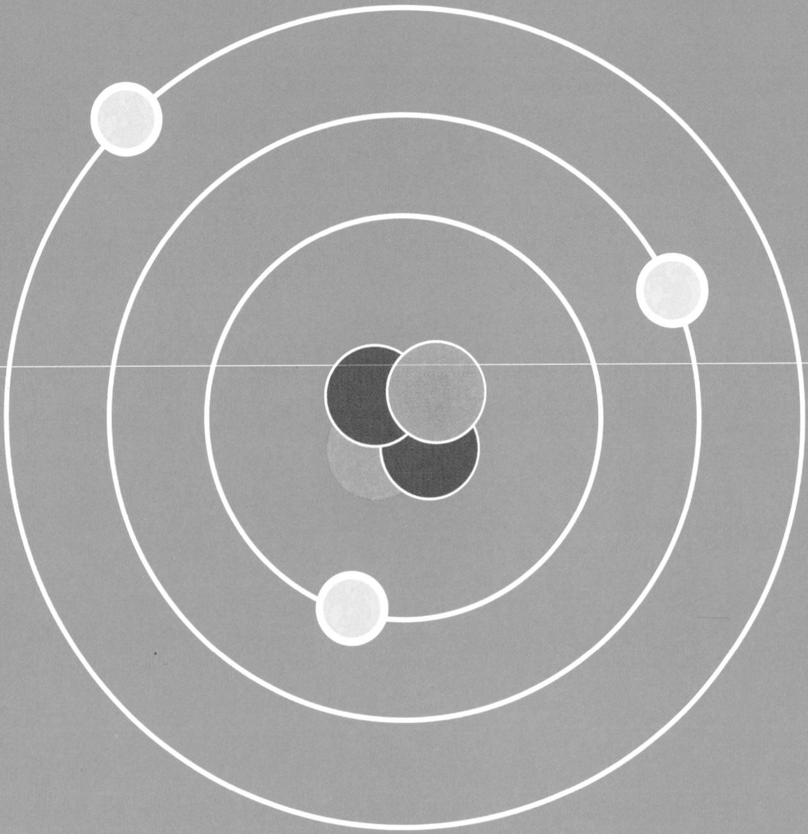

The third part is called
an electron.
It does not live in the middle.

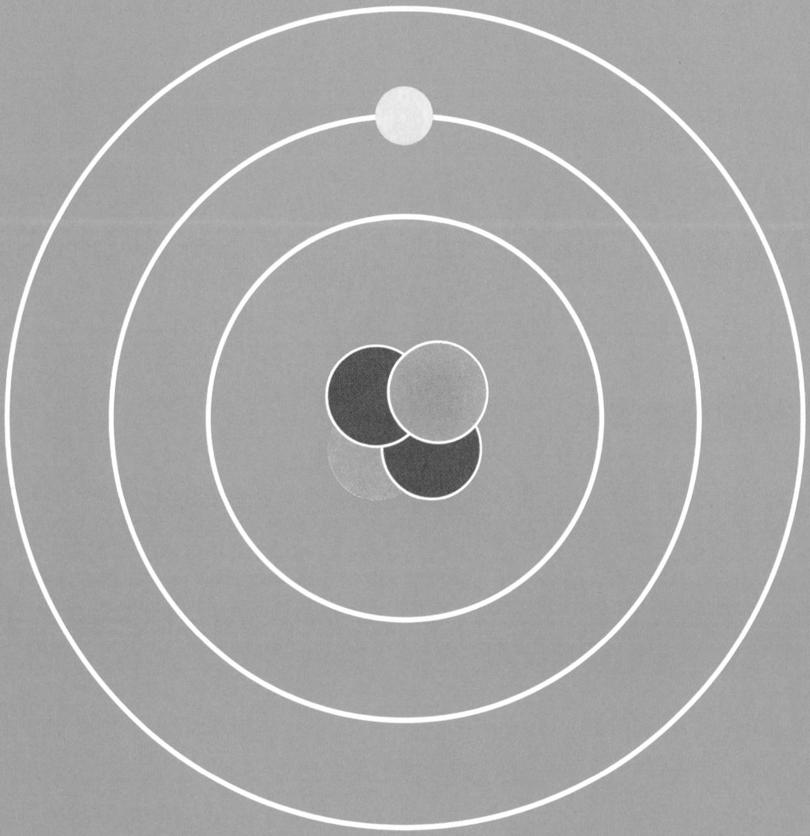

It lives on a ring.

Atoms have lots of rings.

They go around the middle.

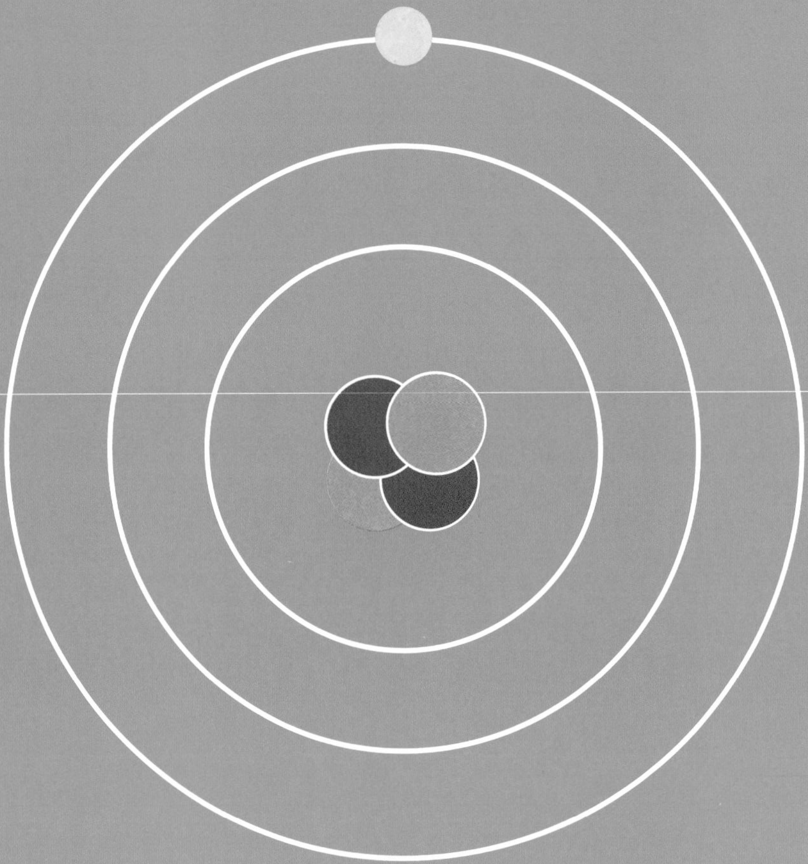

An electron is like a bouncy ball. It can jump up to this ring.

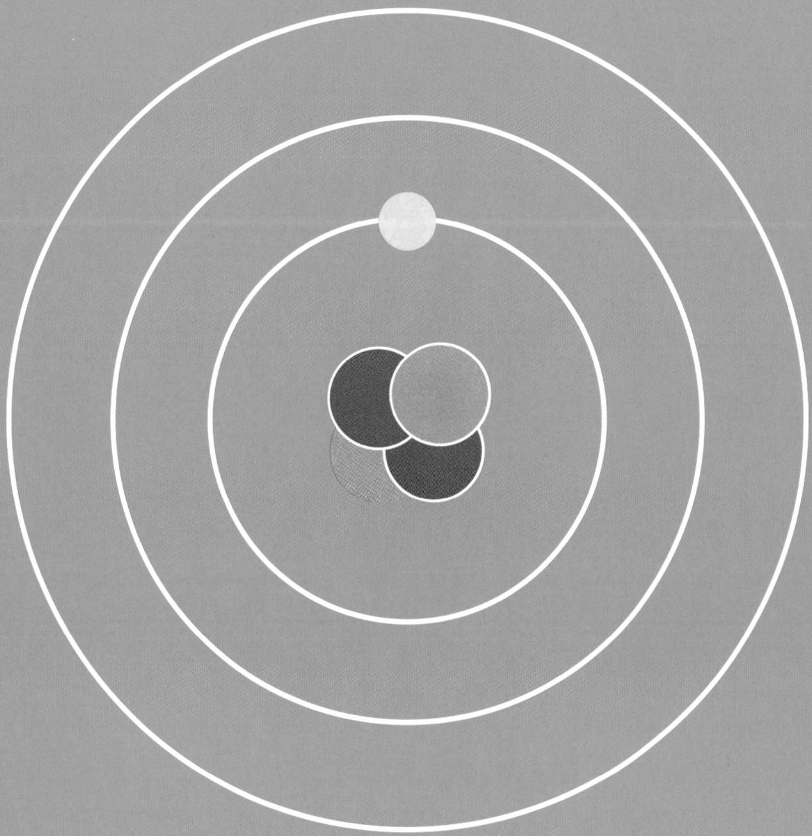

It can fall down
to this ring.

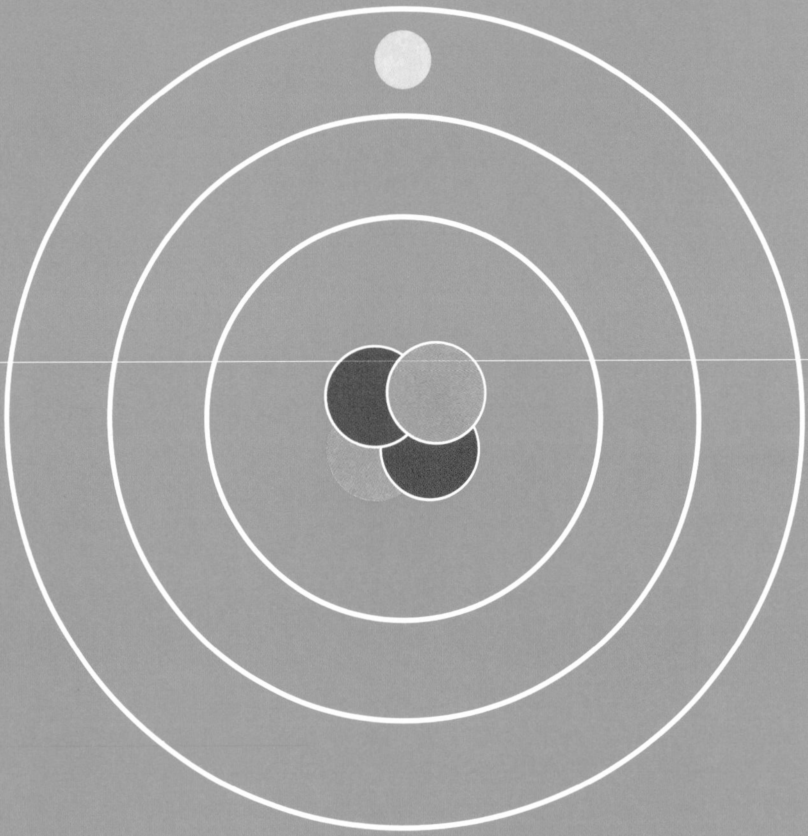

But the ball cannot stay here.
It cannot stay
between two rings.

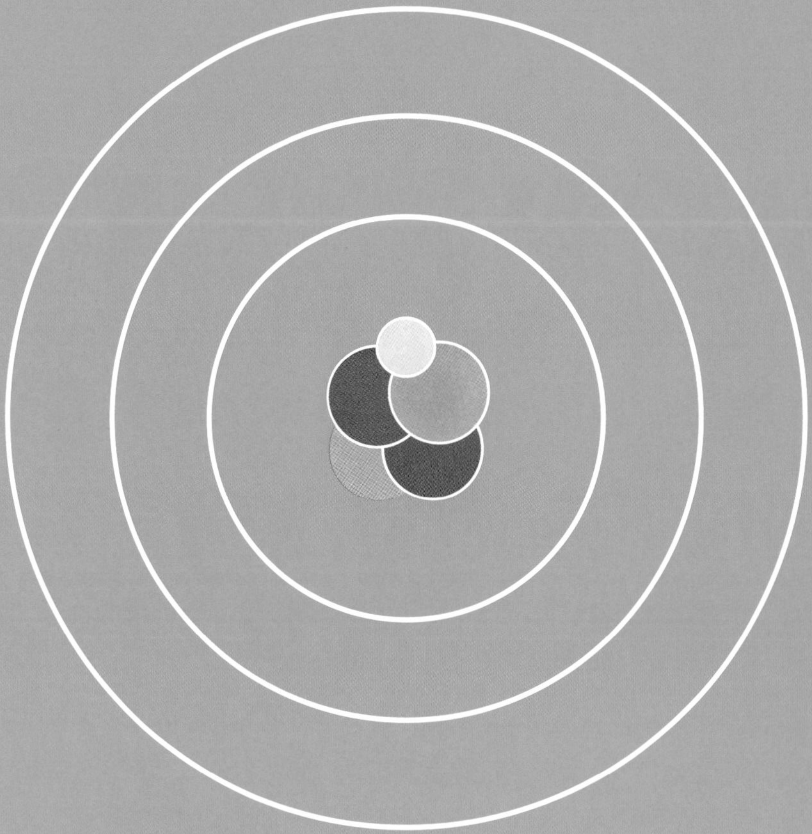

The ball cannot stay
in the middle.

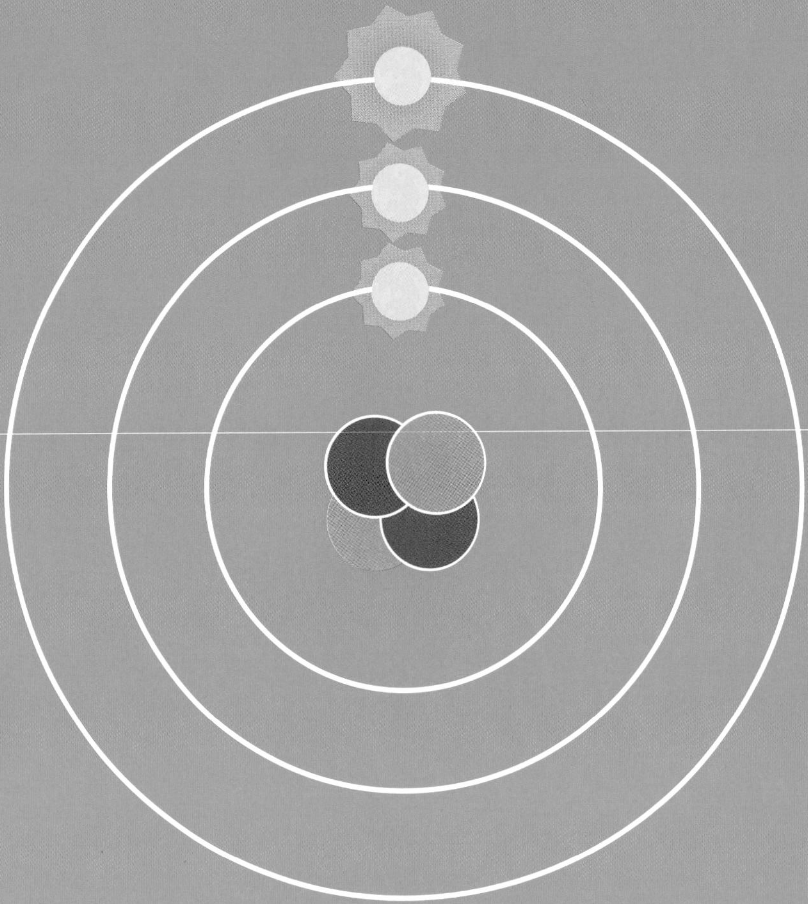

The ball jumps up and down.
It has energy.
But its energy does not stay the same.

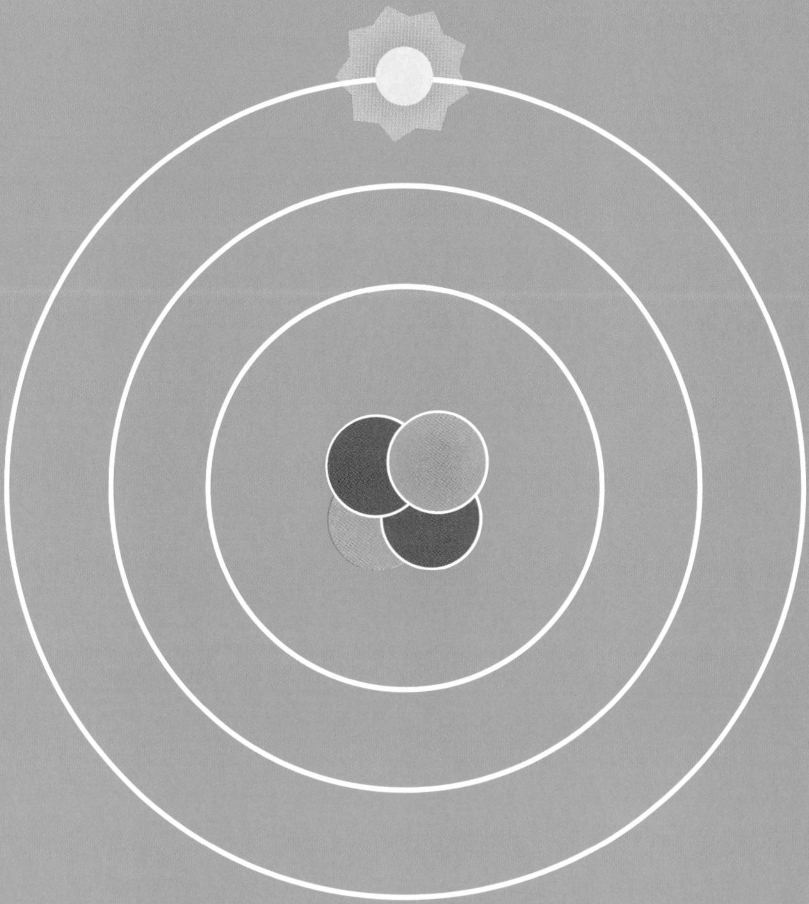

This ball is far from the middle.
It has the most energy.

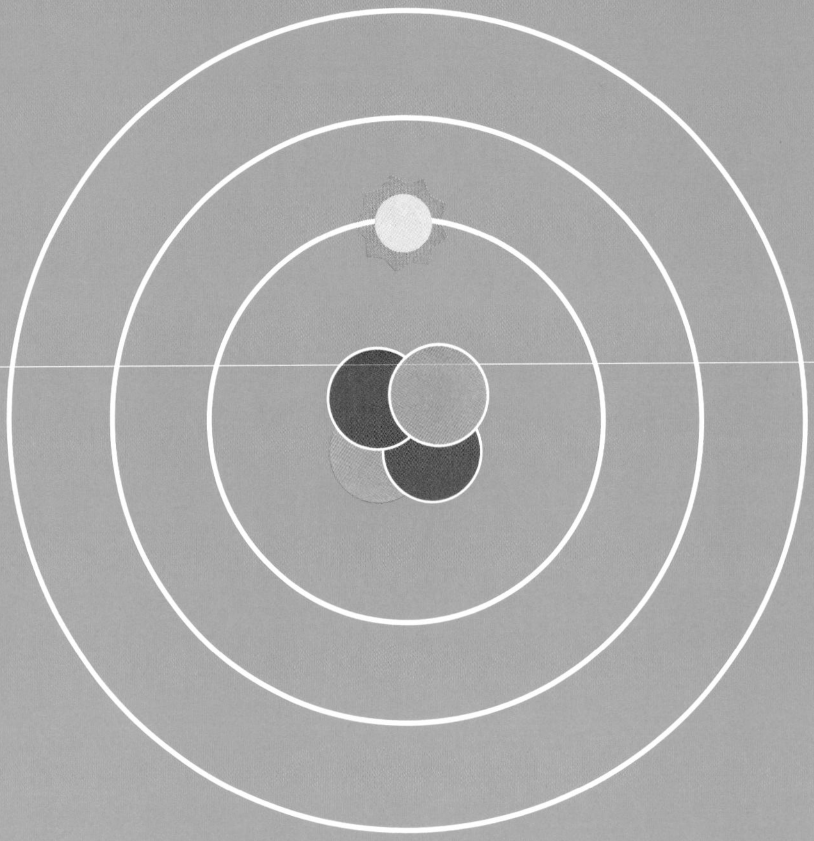

This ball is near the middle.
It has the least energy.

The ball does not stop moving.
It always has energy.

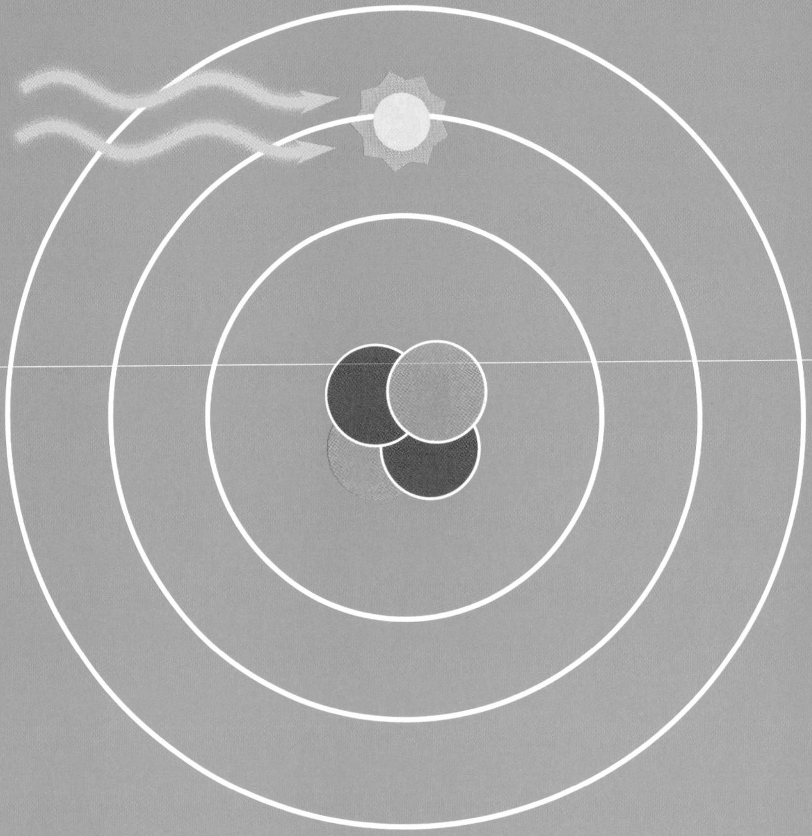

The ball gets more energy when it jumps up.

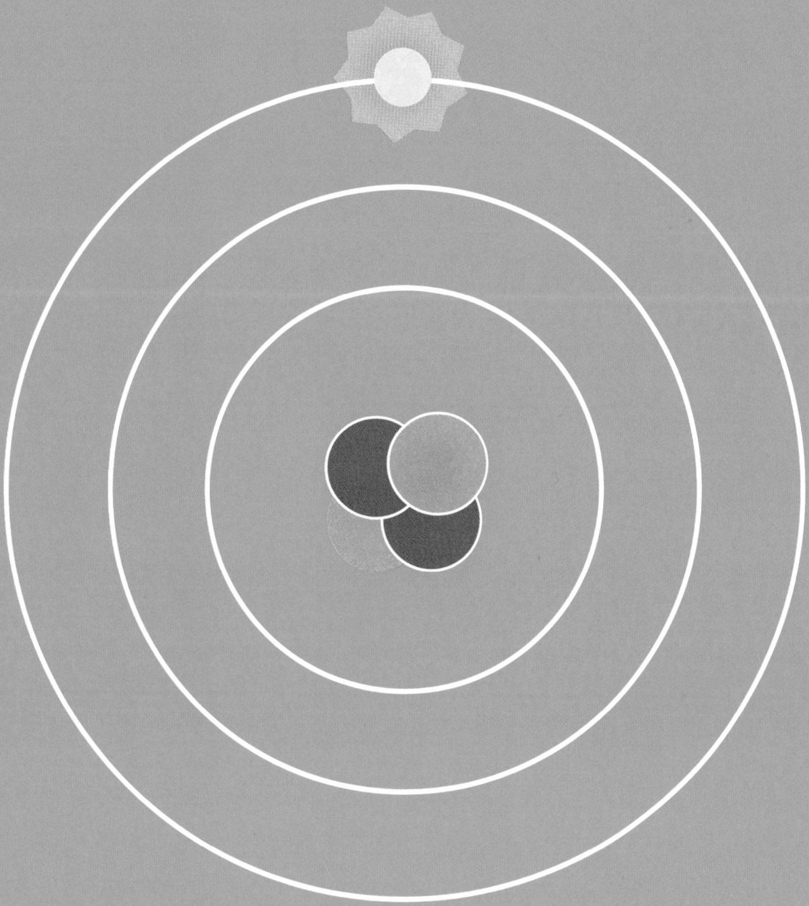

It has more energy
than it did before.

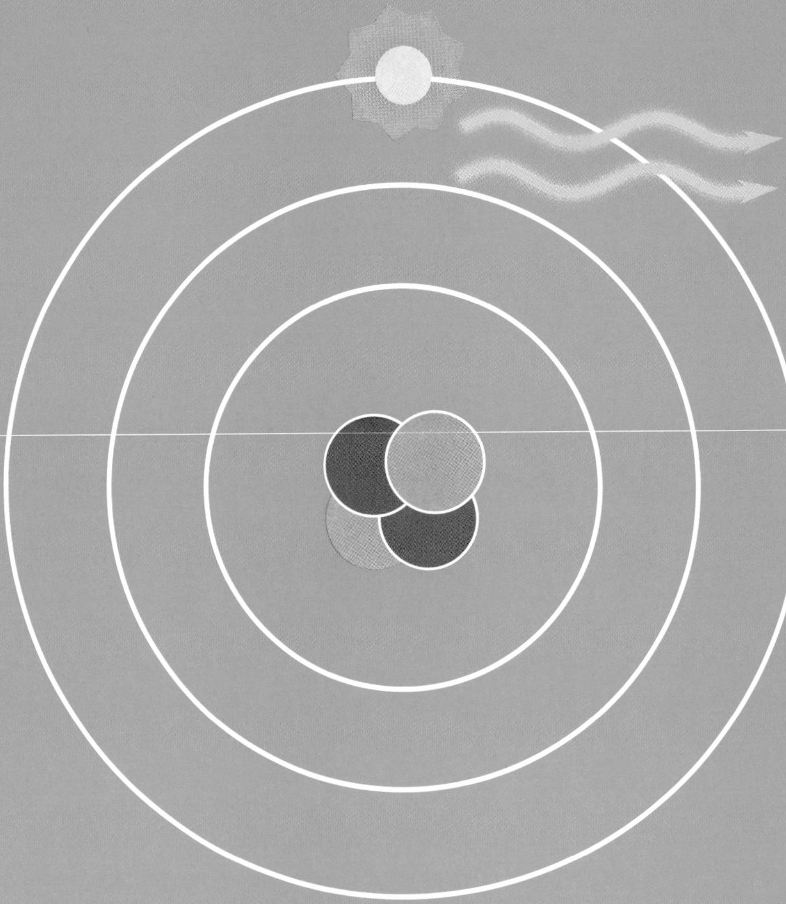

The ball loses energy
when it falls down.

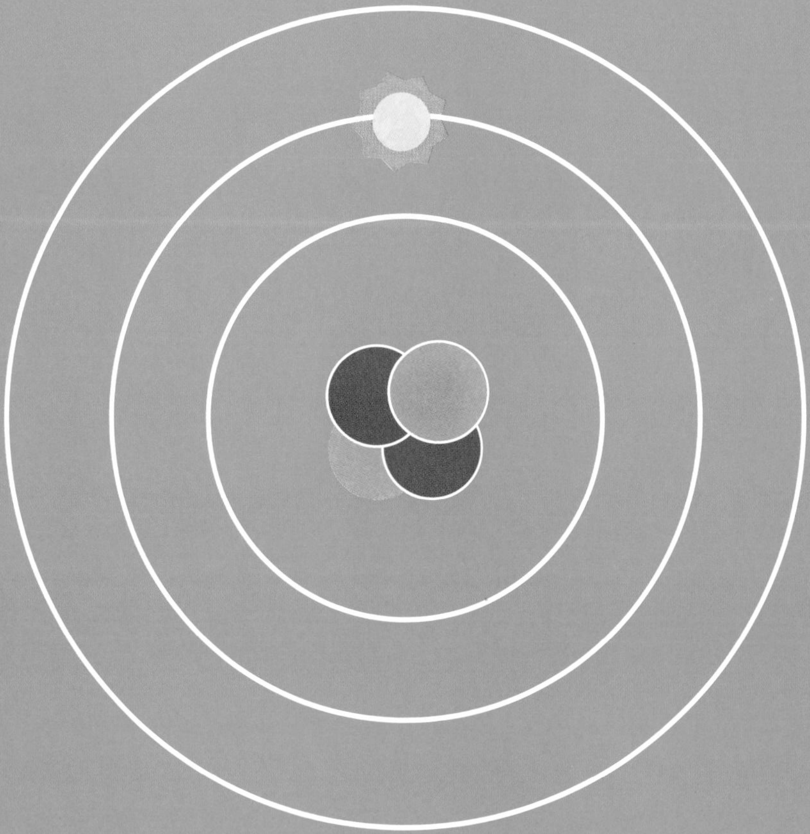

It has less energy
than it did before.

The ball has energy.
Energy can be measured.

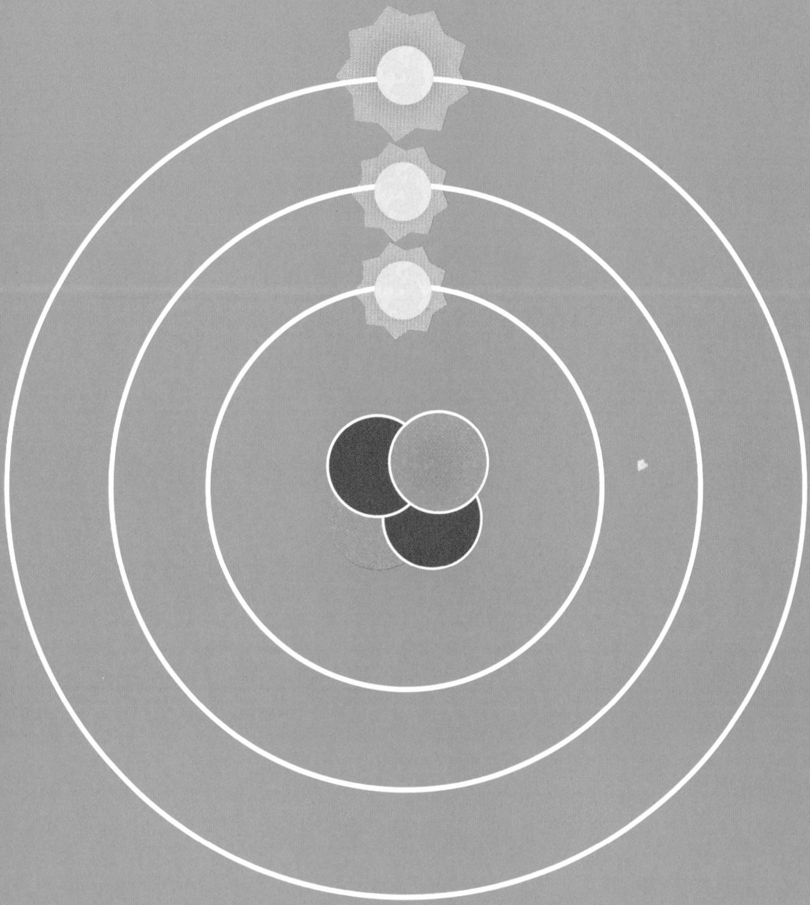

Energy is counted in quanta. Quanta are chunks of energy.

Good job, big thinker.

Now you know
QUANTUM
PHYSICS!